< GET CODING! >

Get Coding with Scratch™

By Jenna Vale

Gareth Stevens
PUBLISHING

Please visit our website, www.garethstevens.com. For a free color catalog of all our high-quality books, call toll free 1-800-542-2595 or fax 1-877-542-2596.

Cataloging-in-Publication Data
Names: Vale, Jenna.
Title: Get coding with Scratch / Jenna Vale.
Description: New York : Gareth Stevens Publishing, 2024. | Series: Get coding! | Includes glossary and index.
Identifiers: ISBN 9781538288726 (pbk.) | ISBN 9781538288733 (library bound) | ISBN 9781538288740 (ebook)
Subjects: LCSH: Scratch (Computer program language)–Juvenile literature. | Microcomputers–Programming–Juvenile literature.
Classification: LCC QA76.73.S345 V35 2024 | DDC 005.26'2–dc23

Published in 2024 by
Gareth Stevens Publishing
2544 Clinton Street
Buffalo, NY 14224

Portions of this work were originally authored by Patricia Harris and published as *Understanding Coding with Scratch*. All new material in this edition authored by Jenna Vale.

Designer: Leslie Taylor
Editor: Megan Kellerman

Photo credits: Cover (illustrations) StonePictures/Shuttertstock.com; cover & series art (coding background) BEST-BACKGROUNDS/Shutterstock.com, (various Scratch screenshots and sprites) Scratch is developed by the Lifelong Kindergarten Group at the MIT Media Lab. See http://scratch.mit.edu; p. 5 Max kegfire/Shutterstock.com; p. 9 Blend Images/Shutterstock.com; p. 11 Veja/Shutterstock.com; p. 13 Prostock-studio/Shutterstock.com; p. 19 wavebreakmedia/Shutterstock.com; p. 27 Tyler Olson/Shuttetstock.com; p. 29 Ground Picture/Shutterstock.com.

Printed in the United States of America

CPSIA compliance information: Batch #CS24GS: For further information contact Gareth Stevens at 1-800-542-2595.

CONTENTS

Words in the glossary appear in **bold** the first time they are used in the text.

Not Just for Scientists

For much of computer science history, using a computer meant typing in commands to make and run the software, or programs. As computers became more advanced and more available, computer scientists worked to make them easier for everyone to use, even if they didn't have any experience. Today, we use computers so often it's hard to believe they were once used only by scientists!

Many computer scientists also work to make coding easier for beginners to learn. The **programming language** Scratch, which is free and maintained by the Scratch Foundation, was made to help kids like you get coding.

You can make a free account to use Scratch right in your web **browser**.

Creating Scratch

The creators of Scratch made a language that allowed coders to mix many different kinds of pictures, sounds, and more all together. They thought it was similar to how a DJ scratches records to make new sounds—that's where the name Scratch comes from!

Text or Graphics

Many coding languages are text-based, or word-based. This means coders must write out the commands they want to give the computer. Coders must understand the correct syntax, or order, in which to put every word and character.

With Scratch, coders don't have to worry about learning every part of the syntax. Commands are packaged in colorful blocks that users drag and drop into a workspace. Scratch uses a graphical user interface, or GUI (GOO-ee). "Graphical" refers to graphics, which are pictures and shapes. "Interface" is another way to describe when two things—such as a computer and a user—communicate with each other.

This is the home page for Scratch. You can view featured projects or start a new coding project right away.

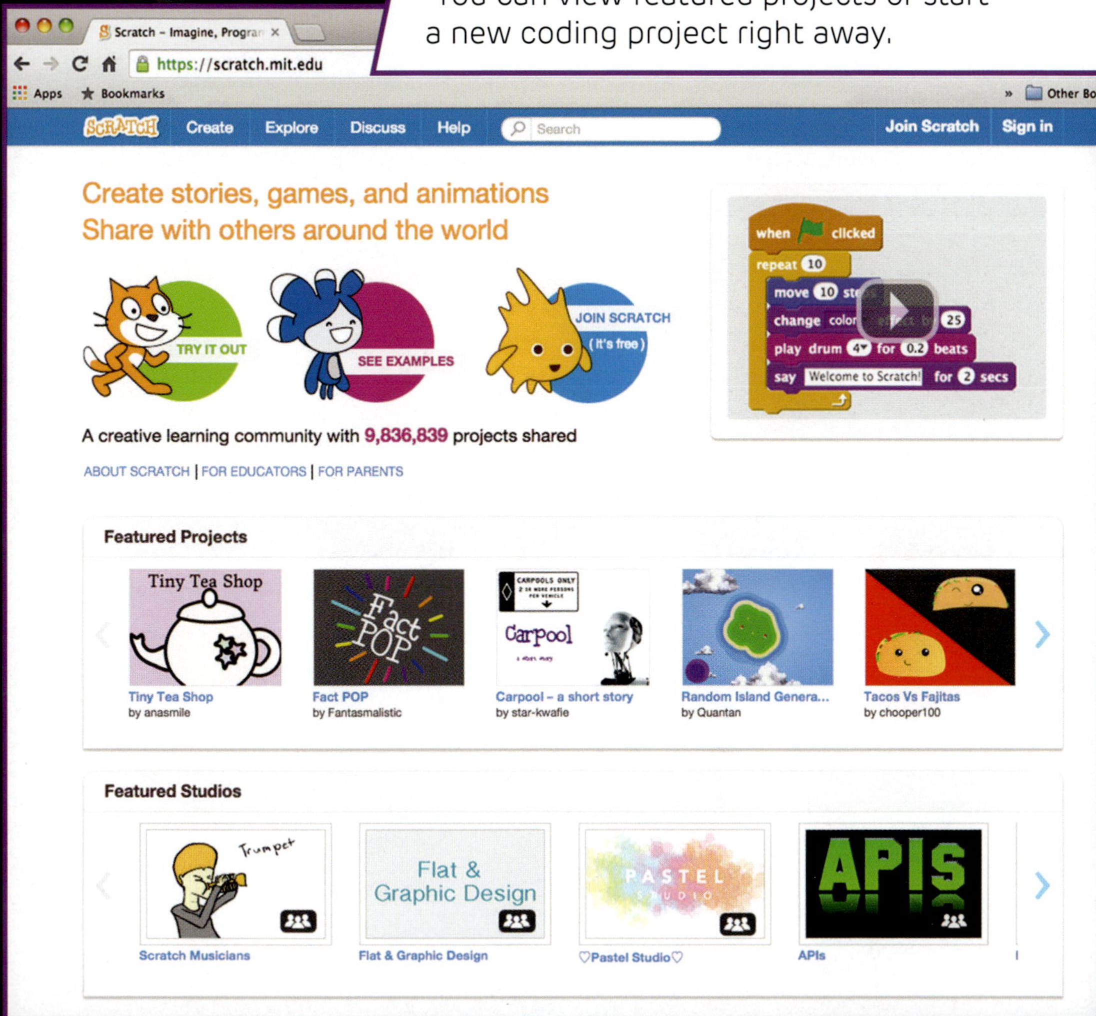

GUIs Everywhere

Before GUIs, interacting with a computer was done through text commands. The first GUIs came out in the early 1970s, and they're very common today. Windows and Apple computer screens are GUIs that make it easy to get to **applications**, or apps—and those apps have GUIs as well.

Coding Basics

Before you start coding, it's important to understand the rules of computer programming. Following these rules will help you code in any language, including Scratch.

Rule 1: Coders must know what they want the computer to do and write a plan.

Rule 2: Coders must use special words to have the computer take input, make choices, and act.

Rule 3: Coders need to think about what tasks can be put into a group.

Rule 4: Coders need to employ **logic** with AND, OR, NOT, and other logic statements as key words.

Rule 5: Coders must explore the **environment** and understand how it works.

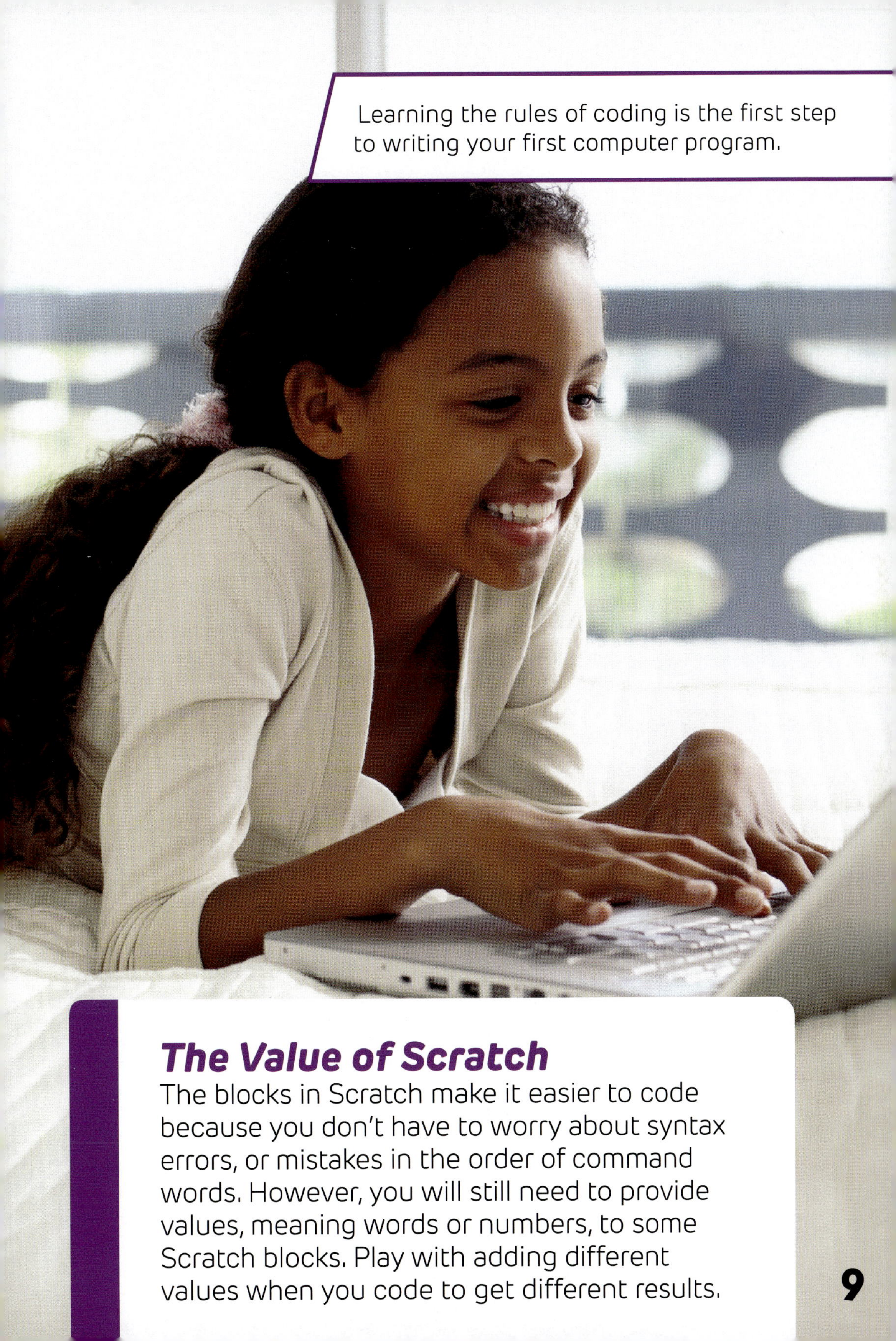

Learning the rules of coding is the first step to writing your first computer program.

The Value of Scratch

The blocks in Scratch make it easier to code because you don't have to worry about syntax errors, or mistakes in the order of command words. However, you will still need to provide values, meaning words or numbers, to some Scratch blocks. Play with adding different values when you code to get different results.

Make a Plan

One of the main types of projects you can do with Scratch is create an **animation**. You choose a stage or background for the action. You choose characters called sprites. You choose movements, sounds, and speech as well. You can even make your own backgrounds, sprites, and sounds.

To begin, make sure you have a plan. Think about input, choices, and actions. Figure out which commands can be grouped. If you want a cat to walk across the screen and meow, you need to find a cat sprite, put it on a stage, and add a meow sound.

These are just a few of the costumes you can choose for your sprite in Scratch.

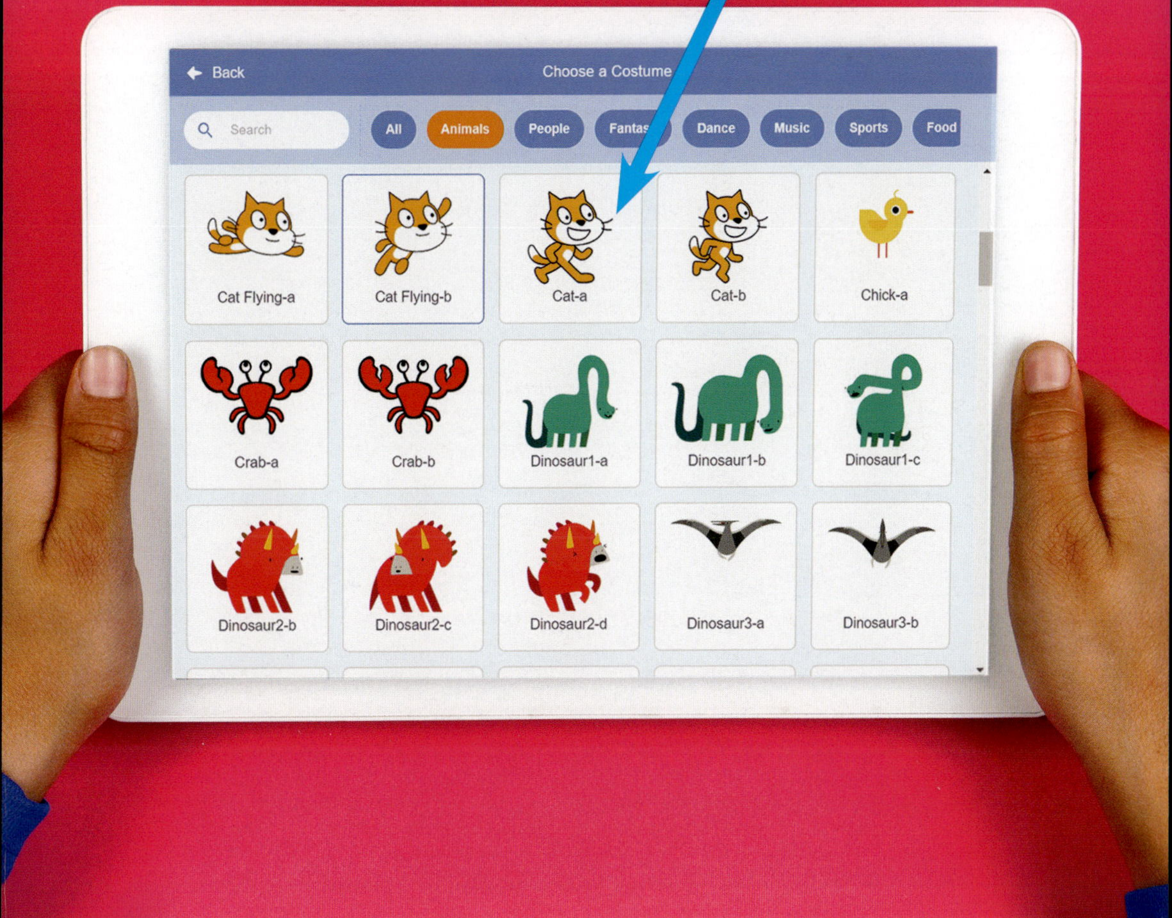

A New Look

Some sprites have more than one costume, which means you can change what they look like. In the top left of the workspace, there's a costumes tab. Select it to make any changes you want to the current sprite, select a new costume, or choose "Surprise" to get a surprise costume. Turn the page to learn more.

your own, or upload a background from your computer.

Changing Scenes

If you plan to make a game or animated story with Scratch, you'll probably want to add scene changes to make things more exciting! Choose all the backgrounds you want to include in your project and make sure they all have **distinct** names. Then you'll be able to use and keep track of them while coding.

Be as creative as you want with your sprites and backgrounds!

In the Workspace

Explore the workspace before you begin your project. The coding block menu is on the left side of the screen. The workspace where you'll drag and drop coding blocks is in the middle of the screen, and the stage, or animation window, is on the right. See the picture on the next page to learn more.

The stage lets you see how your code will work in real time. You can try out many commands by clicking on the coding blocks in the menu. The sprite on the right will do the action.

This is a new project screen in Scratch. The block menu is on the left, the workspace is in the middle, and the stage is on the right.

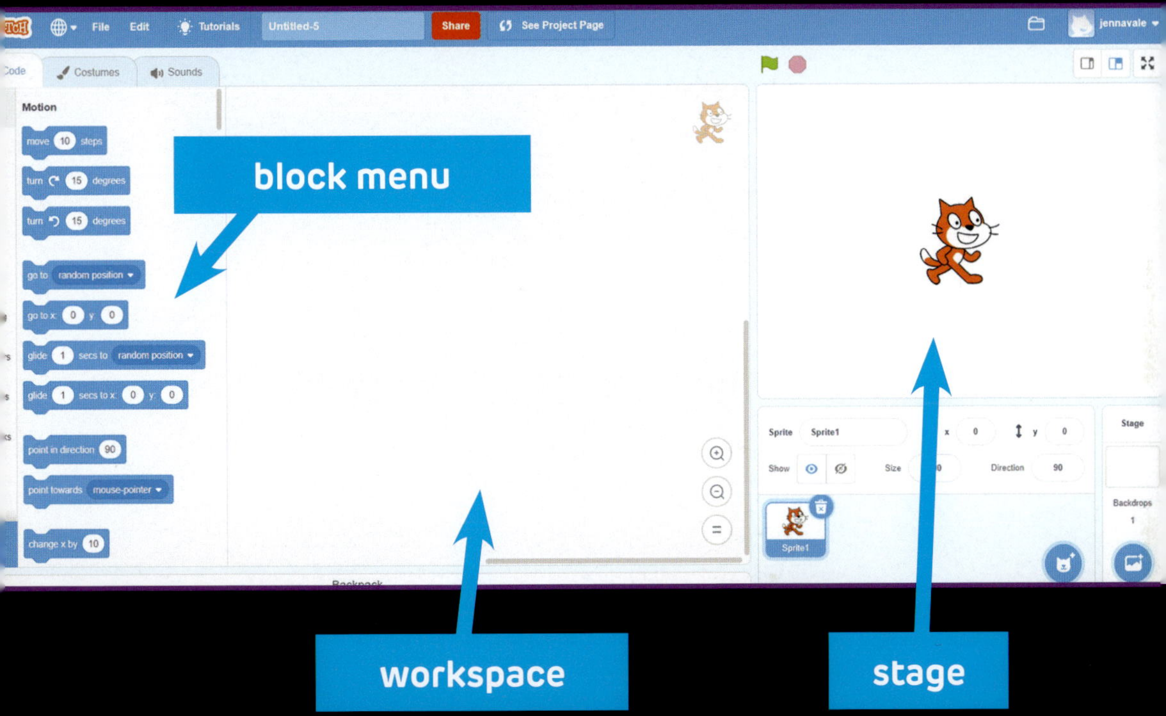

Trial and Error

If you want to see how different Motion blocks work before you begin coding, click on them in the block menu to see how the sprite moves with that code. When you're ready to start coding, you can drag the sprite back to the starting position on the stage.

Sorting Blocks

Pay attention to the kinds of blocks in the menu. What a block does depends on its shape, color, and the text written inside it. Near the bottom of the menu, you might even notice the My Blocks category, or group—this is where you can make your own special blocks to code with.

If you're not sure if certain blocks can be connected, drag them into the workspace and see if their shapes can fit together. Event blocks, for example, are shaped like hats because they are always the top block in a series of coding blocks.

Add an Extension

You can also add Extension blocks to your code menu for certain projects. Select the icon at the bottom left of the menu to add blocks for features such as music or text to speech. If you have a Raspberry Pi, LEGO MINDSTORMS robot, or LEGO WeDo creation, you can even add command blocks specific to those devices.

These are some of the types of blocks you'll use when coding your project.

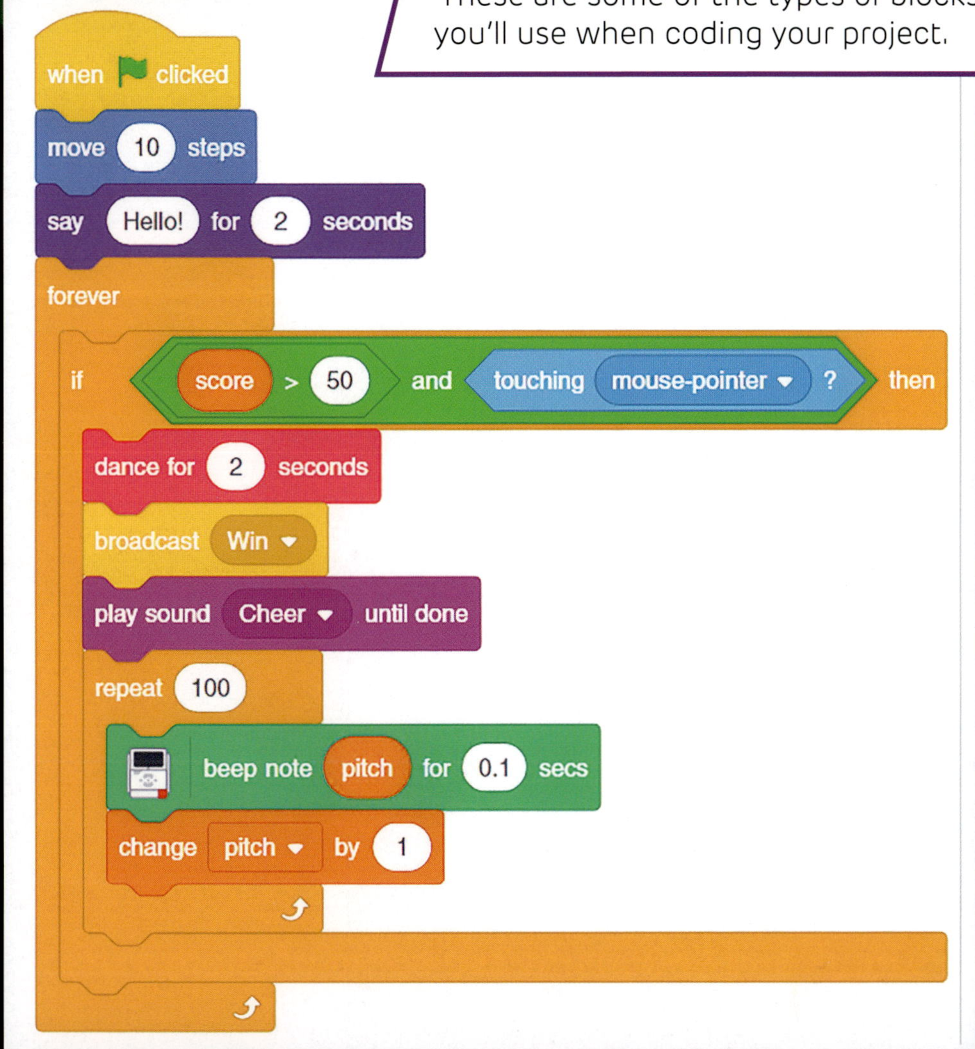

- **Events block:** a gold block that decides what events occur in an animation
- **Motion block:** a blue block that tells how and where the sprite should move Motion blocks have a special shape that shows they can be stacked together
- **Looks block:** a purple block that controls a sprite's appearance
- **Control block:** a yellow block that controls scripts, or stacks of connected blocks. Stacks of blocks can be placed inside C-shaped Control blocks to create **loops**.
- **Operator block:** a green block that performs string handling or math functions
- **Sensing block:** a light blue block that detects something, such as input provided by a user. This block can come in many shapes depending on its use
- **Sound block:** a magenta block used to control sounds

Two other kinds of blocks you'll use can't be stacked because they're just for adding value to other blocks. This means they only go inside of stacking blocks. You'll find these blocks in the Sensing, Operators, and Variables sections of the block menu.

The first kind is called a Reporter block. It's shaped like an oval and reports the value that you add to it. The second kind is called a Boolean block. It's shaped like a hexagon and holds "true" and "false" values. These blocks are more advanced, but pay attention to the examples in this book and you might see some of them in use!

Programming Strings

When you put text into a Reporter block, you're creating a string. In programming, a string is any text that is meant to be read by the computer. In Scratch, this means that any text you have your sprite think or say is a string.

The more you learn about coding blocks, the more you'll be able to code with Scratch.

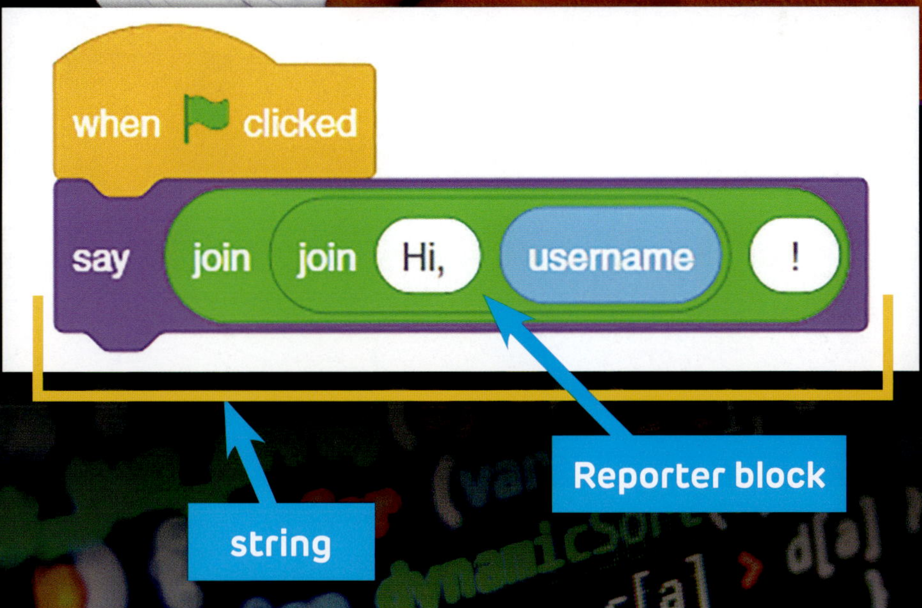

Drag and Drop

Once you're comfortable with the workspace, it's time to get started. For this first project, let's try making a cat walk across a park scene. Use an Events block to start the program.

The first Motion block puts the cat in a good place to start. The yellow Control repeat block then groups tasks together to create an animation loop. The purple Looks blocks change from one cat to another so it looks like it's moving its feet. The blue Motion blocks let the cat move. The Control blocks that say "wait" slow down the computer so you can see the action.

Here's an example of the code for getting the cat to walk across the park scene.

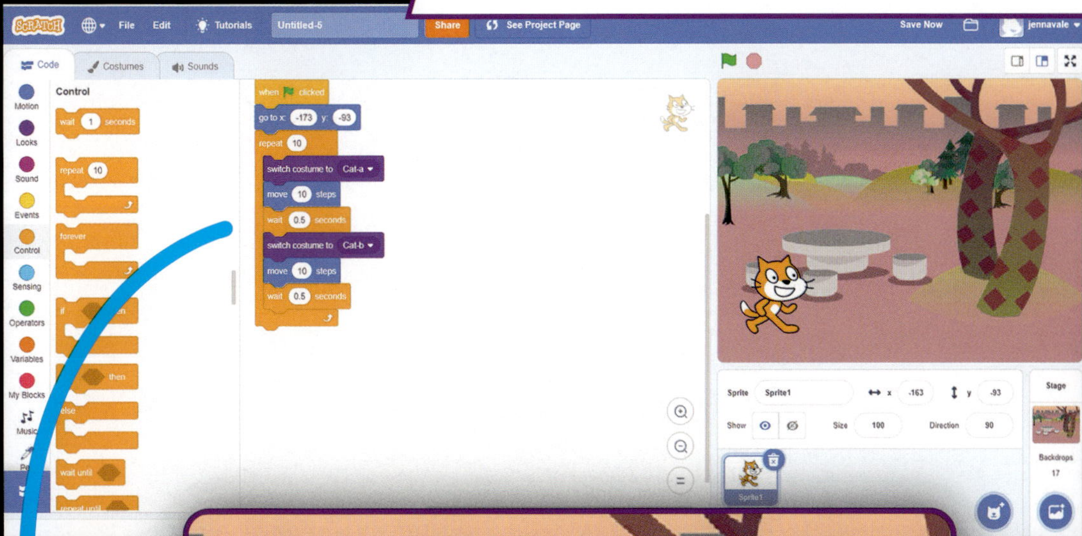

Wait! What Was That?

When you're making a Scratch animation, make sure to add steps to slow things down at first. Computers work so fast that sometimes we can't see what's happening in an animation. We must stop the action for a short time so we can see the changes being made.

If the user says yes, it should meow. Writing this code will include using logic.

This set of blocks lets the computer get some input. It also lets the computer decide what action to take. You let the cat ask a question. The program waits for an answer to be typed. Notice the yellow Control box that says "If." It has a logic statement in it that says if the typed answer is "yes," then the action the cat displays is a repeated meow.

More on Loops

Both code exercises we've done so far feature loops. Grouping commands into repeating loops makes the code shorter. This lets you keep coding without the workspace getting too messy. You can copy and reuse groups of commands in a Control block, or even add them to a new program. That makes coding quicker.

This is just one example of code you can write for your sprite. What else would you code?

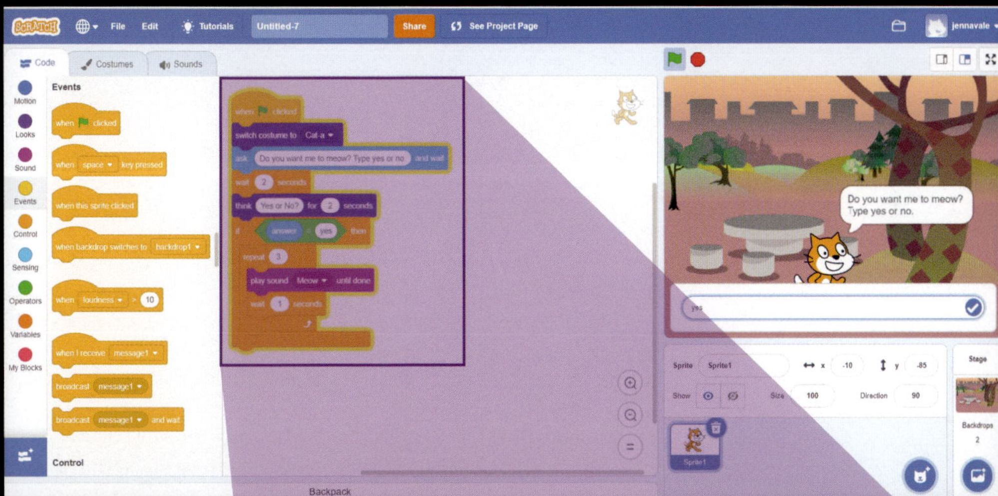

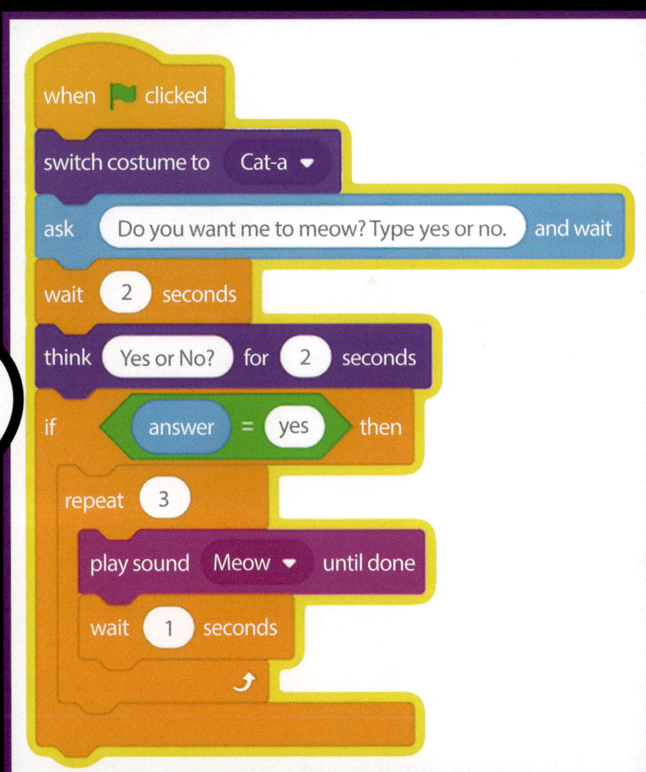

Ready to Code

Once you've had some practice and are ready to create your own animation, remember to follow these steps:

1: Write out a simple plan.
I want a sprite to move across a stage.

2: Choose a stage and a sprite.
I want a cat in a park.

3: Think about actions you can group together so they repeat.
Choose a sprite with costumes. Make the costumes change and the sprite move in a repeat.

4: Think about how to use logic in your program.
I want my cat to walk AND meow.

5: Code!
Have fun with what you've learned and keep going!

A New Adventure

cat code

when [green flag] clicked	The program begins.
say (Hello!)	The cat says hello to the toad.
wait (2) seconds	The cat waits two seconds. Meanwhile, the toad will jump forward.
say (Wait! Where are you going?)	The cat asks where the toad is going one second after the toad appears to jump away.

toad code

when [green flag] clicked	The program begins.
wait (1) seconds	The toad waits one second while the cat says "Hello!"
switch costume to (wizard-toad-b ▾)	The toad switches costumes and appears to run away.

In this code, the cat has run into a mysterious figure on their way to a distant castle. What would you code next in this story?

Sprite by Sprite

One way to test your skills is to create a scene that has more than one character. See the code above for an example of a scene where the cat meets a fellow traveler. Even though both sprites are in the scene, you must code each sprite separately. Pay attention to the timing of each sprite's actions!

Learn More with Scratch

If you have an idea for a project but aren't sure how to make it, you can always find a lesson to help you out. Scratch's tutorials, or guided lessons, can teach you how to code different kinds of projects. Just select "Tutorials" next to the light bulb icon at the top of your project screen.

One of the best parts of coding is learning from others! Scratch is a global community, and you can share your projects with other Scratchers as well. Select "See Project Page" at the top of your project screen to learn more about sharing your coding projects.

Scratch is open-source software, which means it's free for coders to learn with, use, and even improve the program!

ScratchJr and WeScratch

Other ways to learn with Scratch include ScratchJr and WeScratch. ScratchJr is a simpler **version** of Scratch. If you want to spend time practicing the basics, ScratchJr might be a good fit. WeScratch is a series of free online workshops you can do with a parent or teacher. WeScratch workshops are similar to Scratch's tutorials.

Scratch On!

Scratch is now the biggest coding community in the world for beginning coders, but it didn't start out that way. Scratch was very different when it first came out in 2007! Many computer scientists and coders worked together to make Scratch better over the years. Today, with Scratch 3.0, there are so many ways to learn coding and create with Scratch.

Don't forget, the creators of Scratch were beginners just like you once. They got to where they are after a lot of learning, practicing, and having fun. You can do the same by coding with Scratch. It's time to get started. Scratch on!

Go to www.scratch.mit.edu to get started with a free Scratch account.

Get Inspired

No matter what you're interested in, you can draw inspiration from it for coding. One inspiration for Scratch was a simple toy that children had been playing with for years—LEGO building blocks. Blocks of Scratch code are made to snap together just like toy blocks.

GLOSSARY

animation: Drawings or computer images that appear to move.

application: A software program, usually for a mobile device, that serves a specific purpose.

browser: A program that is used to look at websites through the internet.

distinct: Easy to recognize.

environment: The combination of computer hardware and software that allows a user to perform various tasks.

logic: A proper or reasonable way of thinking about or understanding something.

loop: A process that starts over once it reaches the end.

option: A choice.

programming language: A coding language designed to give instructions to a computer.

upload: To move or copy a file from one program to another using the internet.

version: A form of something that is different from the ones that came before it.

FOR MORE INFORMATION

Books

González, Echo Elise. *Logic in Coding*. Chicago, IL: World Book, 2020.

McCue, Camille. *Getting Started with Coding: Get Creative with Code!* Hoboken, NJ: Wiley, 2020.

Wainewright, Max. *Code Your Own Games! 20 Games to Create with Scratch*. New York, NY: Sterling Children's Books, 2020.

Websites

Code.org
www.code.org
An educational nonprofit focused on providing computer science tools for grades K–12.

Scratch
www.scratch.mit.edu
This is the official website for the Scratch programming language.

ScratchJr
www.scratchjr.org
This is the official website for ScratchJr, a simpler version of Scratch to learn with.

Publisher's note to educators and parents: *Our editors have carefully reviewed these websites to ensure that they are suitable for students. Many websites change frequently, however, and we cannot guarantee that a site's future contents will continue to meet our high standards of quality and educational value. Be advised that students should be closely supervised whenever they access the internet.*

INDEX